T0262733

Published by Clanrye International,
55 Van Reypen Street,
Jersey City, NJ 07306, USA
www.clanryeinternational.com

Optoelectronics Handbook
Edited by Rodney Lappin

International Standard Book Number: 978-1-63240-404-6 (Hardback)

Printed in the United States of America.

Optoelectronics Handbook

Edited by **Rodney Lappin**

LANRYE
INTERNATIONAL

New Jersey